A POCKETFUL OF E's

Daphne Metland

A POCKETFUL OF E's
The Pocket Identification Guide to Supermarket Food Label Codes

W. Foulsham & Co. Ltd.
London · New York · Toronto
Cape Town · Sydney

W. Foulsham & Company Limited
Yeovil Road, Slough, Berkshire, SL1 4JH

ISBN 0-572-01366-3

Copyright © 1986 Daphne Metland

All rights reserved.
The Copyright Act (1956) prohibits (subject to certain very limited exceptions) the making of copies of any copyright work or of a substantial part of such a work, including the making of copies by photocopying or similar process. Written permission to make a copy or copies must therefore normally be obtained from the publisher in advance. It is advisable also to consult the publisher if in any doubt as to the legality of any copying which is to be undertaken.

Photoset in Great Britain by Gecko Limited, Bicester, Oxon and printed in Great Britain by St Edmundsbury Press, Bury St Edmunds, Suffolk.

Contents

Acknowledgements/6

Foreword/9

Introduction/10

1. **Shopping Today**/12
2. **The Law and Labels**/26
3. **Additives**/38
4. **Nutritional Labelling**/68
5. **Food Symbols**/76
6. **Glossary of Food Terminology**/78
7. **Finding Out More**/84

   Appendix

Acknowledgements

The information contained in this book has come from many sources. Acknowledgement is due to the following:

Food Labelling Regulations 1984 (No. 1305)

The Vegetarian Society
Parkdale, Dunham Road, Altrincham, Cheshire.

The Vegan Society
33/35 George Street, Oxford, OX1 2AY.

The Hyperactive Children's Support Group
c/o Sally Bunday, 59 Meadowside, Angmering, Sussex, BN16 4BW.

The Ministry of Agriculture, Fisheries and Food
Draft guidelines on nutritional labelling.
Look at the Label
Publications Unit, Lion House, Alnwick, Northumberland, NE66 2PF.

The Soil Association
Look Again at the Label
Walnut Tree Manor, Haughley, Stowmarket, Suffolk, P14 3RS.

Safeways Nutritional Advisory Dept.
Beddow Way, Aylesford, Maidstone, Kent, ME20 7AT.

British Home Stores
Food Technology Dept., Marylebone House, 129 Marylebone Road, London, NW1 5QD.

Marks and Spencer
Information Office, Michael House, Baker Street, London, W1A 1DN.

Tesco
Information Office, Tesco House, PO Box 18, Delamere Road, Cheshunt, Waltham Cross, Herts, EN18 9SL.

Waitrose
Information Office, John Lewis Partnership, 4, Old Cavendish Street, London, W1A 1EX.

Presto
Information Office, Argyll Stores, Argyll House, Millington Road, Hayes, Middlesex,

Sainsburys
Stamford House, Stamford Street, London, SE1 9LL.

The Coeliac Society Medical Advisor
Coeliac Society of UK, PO Box 181, London, NW2 2QY.

The Health Education Council
78 New Oxford Street, London, WC1A 1AH.

London Food Commission
PO Box 291, London, N5 1DU.

Rita Greer
Bunterbird, 225 Putney Bridge Road, London, SW15 2PY.

The Coronary Prevention Group
60, Great Ormond Street, London, WC1N 3HR.

BBC TV
Food and Drink, BBC TV, London.

Consumers Association
A Research Study on Behalf of the Consumers Association, the Ministry of Agriculture, Fisheries and Food, and of the National Consumer Council: Consumers Attitudes to and Understanding of Nutritional Labelling
14 Buckingham Street, London, WC2N 6DS.

Food and Drink Federation
6, Catherine Street, London, WC2B 5JJ.

The National Eczema Society
Tavistock House, Tavistock Square, London, WC1H 9SR.

The Asthma Society
300, Upper Street, London, N1 2TU.

Action Against Allergy
43, The Downs, London, SW2 8HG.

Ecoropa
Food Additives – Are the Risks Worthwhile?
Ecoropa, Crickhowell, Powys, Wales, NP8 1TA.

Foreword

This book could change your life. If, like so many of us, you've come to care about what's *in* the food we eat – if you're worried about whether all of it's good for us – this book has the answers.

'E' numbers have become a kind of bogeyman to most ordinary people. One 'E' number on a packet is a worry – three is trouble – five is danger. Yet many 'E' numbers are valuable, helpful and harmless. They keep food fresh, bacteria-free and safe for us and our children.

The trick is to be able to tell the difference between the dangerous additives or colours – substances we don't need or want – and preservatives or emulsifiers that we should be grateful for. Daphne Metland's book does the job – an easy reference to all the 'E' numbers and a guide to supermarkets – which are taking the dangerous substances out of their products and which aren't that interested. It's a book that should be in every shopping trolley. It's going to be in mine from now on!

Michael Barry
Presenter of BBC TV's
Food and Drink.

Introduction

What goes into our food has, for too long, been a mystery. Recent changes in legislation and public demand have begun to change that.

Information and advice is now available from a wide range of sources including pressure groups like the London Food Commission and support groups like the Hyperactive Childrens' Support Group. Chapter 7 lists some organisations I have found very helpful in researching this book.

The supermarket chains are, generally speaking, taking steps to give information to customers through leaflets and advisory panels. Safeways lead the field with their quite comprehensive range of leaflets on additives. Sainsburys, Tesco and the Co-Op stores have excellent leaflets on the general issue of healthy eating.

Whilst reading food labelling regulations and draft guidelines has been demanding, meeting and talking to food technologists and nutritionists as well as representatives from the various organisations mentioned in the book was quite fascinating.

What we eat and what makes up a healthy diet is a large and constantly changing field. This book can only be a handy guide: use it to dip into, check information with, and as a guide to food labels.

All the while there is public interest and demand there will be pressure to improve our food. The Hyperactive Children's Support Group was set up nine years ago. Progress has been slow but now they have over 6,000 members. The group receives over 100 letters every day. They are pleased that, at last, their concerns about food additives are being taken seriously. They are finding that doctors and health visitors are suggesting parents contact them for advice on diet for hyperactive children.

Many thanks must go to Sandra Croker the nutritionist who read and checked the information in this book and to my assistant Alison Moore who diligently toured the supermarkets, note book in hand, and read every food label in sight!

Daphne Metland.

Chapter One
Shopping Today

We are what we eat, or so the saying goes. Indeed the food we eat has a great influence on our health and well being. Over the past couple of decades what we eat, and how much we eat, has changed greatly.

We have a wide choice of foodstuffs in our shops and supermarkets. Exotic fruits and vegetables are flown in from all around the world. Canned, packaged and convenience foods of all types are stacked on every supermarket shelf. The average family could buy complete meals and finished dishes every day if they wanted to, and could afford it. Add to this the increasing trend to eat out. Many caterers use some convenience foods and some even use ready-prepared meals. All this means that the number and type of preservatives, colours, flavours and other additives found in the average diet has increased greatly over the past few years. The responsibility for many illnesses such as hyperactivity in children, allergic responses including rashes, breathing difficulties, streaming eyes, etc., and even kidney stones, ulcerative colitis and possibly some cancers has been attributed to some additives.

However, let's not assume that everything was fine in the 'good old days'. Inadequate food was common; poor diets with monotonous and poor quality food was the lot of most people. The same may be said for many parts of the world now. In medieval times food supplies relied heavily on the seasons and the weather. During the winter there was usually insufficient fodder to keep animals alive. Most were slaughtered in the autumn and the meat salted or dried, not always successfully. Herbs and spices from the East were greatly prized imports because they hid the taste of bad meat! Bakers tended to adulterate flour. Often the baker's oven was the only one in the village so people took their flour to the baker to make bread for them. He sometimes creamed off a little of the flour and replaced it with talc or chalk or even alum to make up the weight! Poor storage of damp corn allowed a fungal growth to develop, which on several occasions poisoned many thousands of people in Europe when it was made up into bread.

Certainly in the past our diet has not seemed to be an important aspect of keeping healthy. Certain foods were sometimes endowed with curative properties; there was once a fashion for drinking vinegar with meals to help society ladies to lose weight! Generally, however, life was pretty violent and short anyway and it mattered more to eat enough than to question what was being eaten.

Gradually as the food supplies of the affluent half of the world have grown to an embarassing surplus (while the other half of the world still does not have enough) we have been in a position to question what goes into our food and what makes a healthy diet. Recent reports have suggested that the average Western diet is in fact rather bad for us. Amidst much political controversy the NACNE (the National Advisory Committee on Nutrition Education) and COMA (the Committee on Medical Aspects of food policy) reports have suggested that quite major changes are needed to help improve the health of the nation. NACNE, in September 1983, came to the conclusion that we should reduce sugar consumption by 10% immediately. Fats should be reduced by 10%. Fibre should be increased by 25%. Different types of fats should be eaten, with polyunsaturated fats replacing some of the saturated fats. (Unsaturated fats are found mainly in vegetable sources, saturated ones are from animal sources.) Salt should be reduced by 10%. They consider it will take around 15 years for these changes to take effect. As anyone who has tried to reform their family diet overnight will know it is much easier to make dietary changes gradually.

There is also a great interest in eating healthily today.

Along with the 'health-and-fitness' boom has come an interest in healthy eating and healthy foods. It is possible for the average family to change their diet but this must be accompanied by changes in manufactured foods.

Since we eat more prepared and packaged foods than ever before, so the manufacturers will have to bring their food in line with healthy eating. 70% of our salt intake comes from prepared foods; only 30% is added in cooking and at the table. This also means using less artificial additives. Colourings, flavourings and preservatives have all been used quite liberally in the past meaning that many of us have been consuming a cocktail of chemicals along with our foods. New ranges using less of these additives and replacing the more worrying ones with natural alternatives are now coming into the shops.

However, it is still up to the shopper to read the label and buy wisely. In researching this book attitudes among supermarkets to additives varied greatly. Some have made efforts to remove additives and issue lists of the additive-free foods which they stock. Others have done little or nothing. The same goes for manufacturers. Some foods have had new recipes and are now available in different forms; fruits are canned in natural juices; fish is coloured with natural colours instead of Brown FK, Tartrazine and Sunset Yellow. In some cases the manufacturers have launched new ranges alongside their old ones giving the customer the choice. So vote with your purse! Read the label and opt to buy what best suits your family!

Criticism of the current system which allows the food industry (in the form of the Food Advisory Committee) to advise the Ministry of Agriculture, Fisheries and Food, and to do so in secret, has come from many quarters. The London Food Commission in 1984 published a list of action to be taken to tackle the additive problem. It included: broadening the membership of the Food Advisory Committee, conducting its work more publicly, government control of the testing of additives, and fully detailed labels with health warnings where necessary.

Until such time as this sort of action is put into effect most of us have to rely on shopping selectively. Read the labels and compare them with the lists in this book.

Choose products with the minimum of additives and avoid the more contentious ones. When it comes to the crunch the food industry will give the consumers what they want if not doing so begins to affect their profits.

The Supermarkets

The supermarkets have been at the forefront of responding to consumer pressure to change policy on additives. Those with large ranges of own label products have been in a particularly strong position. They are able to make changes themselves rather than attempt to pressurise other manufacturers. They come into most contact with the customer and are able to sell a wide range of foods (sugar-free, low-fat or additive-free products) side by side with ordinary ones. Many supermarkets have already begun phasing out additives or are opting for safer alternatives where necessary. Colours are reasonably easy to change. There is a selection of natural colours available. Preservatives take longer. Tests of shelf life of products are necessarily time-consuming, so changes here have been even slower but will begin to make themselves more obvious over the next couple of years.

The interest in healthy eating extends beyond additives though. Many supermarkets (and some companies) are now producing alternative products such as sugar-free jams, which is not quite the impossibility it seems. (Fruits contain natural sugars so the jam can be made entirely of fruit.)

The 'ready-prepared meal' market is growing rapidly. It was worth over £75 million in 1985. Meals sold frozen or chilled, ready to cook at home, were made very popular by shops such as Marks and Spencer and this has been expanded by many supermarkets. Here too there are changes in the type of meals, and what goes into them. British Home Stores are selling a pie filled with a meat substitute called mycoprotein which is high in fibre and low in fat – ideal for vegetarians and for anyone keen on eating less meat.

Hopefully the development of new foods will lead to more and more healthy alternative products and the promotion of long established alternatives like soya bean curd which has the potential to be much more widely used. Ten years ago wholemeal bread was a specialist variety in the supermarkets. Now it is the first type to sell out. Perhaps the same pattern will follow with other 'health' foods.

Boots
Food Marketing Department, Nottingham, NG2 3AA.
Larger Boots stores have food centres which sell a wide range of foods including specialist foods such as gluten-free and those suitable for diabetics. They have their own brand range of vegetarian foods and a new wholemeal bread range. Their 'Second Nature' range of health foods is free from artificial colours, flavours and preservatives. They use nutritional labelling on most foods and around 65% of their own brand lines are free from contentious additives. They are currently working on removing or replacing additives where possible and dicussing this topic with the suppliers of their branded products.

Look out for their 'Dietary Analysis' leaflet in stores which offers a breakdown of your diet into protein, carbohydrate, fat, vitamin and mineral levels. It is a free service run in various stores at various times. Also available in store is their range of leaflets called '*Food Facts*'. They have a Health and Nutrition advisor to answer specific queries on diet.

F O O D
F A C T S

This dish is just one of an exciting range of vegetable and pulse-based meatless dishes, created for Boots. It is now widely accepted that for a healthier way of life, most people would benefit from increasing their vegetable intake. Boots vegetable-based foods add taste, interest and variety to the diet – not just for vegetarians but for everyone who cares about what they eat.

This Country Casserole recipe has been specially created to provide a delicious meal for all the family.

360 gram   12.7 oz ℮

FOR BEST BEFORE ℮

NUTRITION INFORMATION	per 100 gram	per pack
Protein	5.6 gram	20.2 gram
Fat	3.1 gram	11.2 gram
Carbohydrate	15.1 gram	54.4 gram
Dietary Fibre	3.1 gram	11.2 gram
Energy Value	452 kJ	1627 kJ
	107 kcal	385 kcal
	(Calories)	(Calories)

Free from artificial colours, flavours and preservatives

> INGREDIENTS: Water, Potato, Lentils, Peas, Carrot, Celery, Sweetcorn, Red Pepper, Vegetable Oil, Vegetable Stock, Wheatflour, Butter Oil, Caseinate, Potato Starch, Cheese, Salt, Lactose, Natural Flavouring, Thickener (Modified Starch), Mixed Spices, Emulsifier (E340).

Country Casserole – Ready Meal (360g)
Serves 1 or 2

SAINSBURY'S

J Sainsbury PLC
Stamford House, Stamford Street, London, SE1 9LL.
Sainsburys are well on the way to removing many artificial additives: they already have colour-free ice cream, fish fingers and yoghurts free from artificial additives. Brown FK has been removed from their own brand frozen and packaged mackerel and kippers.

They are using nutritional labelling on many products. A free leaflet is available in their supermarkets called *A Healthy Look at Sainsburys' Labels*. They will also advise on special dietary needs. Write to the above address for more information.

INGREDIENTS: PORK (MINIMUM 65%), WATER, RUSKS, POTATO STARCH, SALT, SPICES, DI AND TRI PHOSPHATES, PRESERVATIVE: E221; HERB, ANTIOXIDANT: L-ASCORBIC ACID

Typical Lean 53%, Typical Fat 12%
Contains less than half the fat of a normal pork sausage
Cooking instructions:
Grill for 10 minutes, turning occasionally
J Sainsbury plc Stamford Street London SE1 9LL

NUTRITION	TYPICAL VALUES (GRILLED)	
	PER 100g (3½ oz)	PER SAUSAGE
ENERGY	210 K/CALORIES	100 K/CALORIES
	890 K/JOULES	425 K/JOULES
PROTEIN	15.8g	7.5g
CARBOHYDRATE	12.2g	5.5g
TOTAL FAT	11.5g	5.5g
of which POLYUNSATURATES	1.0g	0.2g
SATURATES	4.8g	2.3g
ADDED SALT	1.8g	0.9g

Pork Sausages (454g)

Safeway

Safeway Food Stores Ltd., Beddow Way, Aylesford, Maidstone, Kent, ME20 7AT.

This chain has announced their plan to remove all contentious additives from their own label products over a period of two years. Already many additive-free items are appearing in the stores and they also run a Nutritional Advisory Service. They deal with letters and requests for information on the contents of their own label brands from shoppers and produce a range of useful leaflets available in the shops or by post. These include *Additives, Nutritional Labelling and Dietary Guidelines.* They also sell a range of organically grown products in many stores. Labelling of specialist food is used in their own label range including 'gluten-free' labelling and the 'growing leaf' symbol (see page 77). All new products carry nutritional information.

Mixed Nuts & Fruit (650g)

Tesco

Tesco House, PO Box 18, Delamare Road, Cheshunt, Waltham Cross, Herts, EN8 9SL.

Tesco launched their healthy eating plan in January 1985. This Leonardo-inspired logo is shown to signify healthy foods. The foods are also described with words such as 'High in Vitamin C' or 'No added sugar' or 'High-fibre'. A range of leaflets are available free in stores covering aspects of healthy eating, and nutritional information is given on own label products.

On the additives question Tesco say they are seriously looking at removing contentious additives from their own label foods. They have made a start with Tartrazine and plan to remove colours, Benzoates (E320 BHA, E321 BHT) and some preservatives. They run a dietary information service and will supply a computerised print-out for Tesco own label products which gives a complete breakdown of the ingredients in each food. Available free from:

Tesco Stores Ltd.
Group Technical Service Centre
43 – 47 High Street, Ware, Herts, SG12 9BA.

NUTRITION

Baked Beans are extremely high in dietary fibre and contain less than 1.5g of fat per serving.

AVERAGE COMPOSITION	PER 140g (5oz) serving	PER 100g (3½oz)
Energy	553kJ/132kcal	394kJ/94kcal
Fat	0.7g	0.5g
Protein	7.0g	5.0g
Carbohydrate	25.2g	18.0g
Fibre	10.2g	7.3g
Added Salt	1.4g	1.0g
Added Sugar (Sucrose)	6.0g	4.3g

INFORMATION

DIRECTIONS
May be served hot or cold. If required hot, empty contents of can into saucepan, heat and stir gently to prevent burning.

INGREDIENTS
Beans, Tomato Purée, Water, Sugar, Salt, Modified Starch, Onion Powder, Spices.

15.5 oz 439 g ℮

Produced in the UK for Tesco Stores Ltd., Cheshunt, Herts.

Beans in Tomato Sauce (439g)

NUTRITION

One Cox's apple provides over one fifth of the daily requirement of Vitamin C which we need to ensure general good health.
They are also a good source of dietary fibre and contain less than 60kcalories each.

1 AVERAGE SIZED APPLE		
AVERAGE COMPOSITION	PER 125g (4½oz) serving	PER 100g (3½oz)
Energy	243kJ/58kcal	193kJ/46kcal
Fat	Trace	Trace
Protein	0.6g	0.5g
Carbohydrate	15g	12g
Fibre	2.5g	2g
MINERALS/ VITAMINS	% RECOMMENDED DAILY AMOUNT	
Vitamin C	21%	5mg

THIS PACK CONTAINS APPROX 7 SERVINGS

INFORMATION

NUTRITION PANEL

All Tesco own-label products will be nutritionally labelled. <u>Above</u> is an example of the type of information which will appear on the product label. In this case it refers to one Cox's apple.

HEALTHY EATING LOGO

On the <u>left</u> are examples of the Healthy Eating Logo which will appear on those Tesco products having a particular benefit which can contribute to a healthier diet.

Tesco Nutritional Information

British Home Stores
Marylebone House, 129, Marylebone Road, London, NW1 5QD.
British Home Stores already have their own range of natural foods. Each new food is being developed with natural additives wherever possible and they are in the process of removing thickening and gelling agents, some colours and monosodium glutamate from many foods. Foods are marked as 'Suitable for vegetarians' where appropriate and they issue lists of British Home Stores' foods suitable for anyone on specialised diets. Write to the above address. Their own nutritional labelling is used on many products.

INGREDIENTS:

Courgette, milk, beef, pasta (durum wheat, spinach), tomato, onion, wheat flour, cheese (cheddar, processed cheddar), hydrogenated vegetable oil, tomato puree, chicken stock, salt, bay, spices, garlic powder.

NUTRITIONAL DATA	AVERAGE PER 100g
ENERGY VALUE	131K cals (460kJ)
PROTEIN	3.4g
FAT	7.7g
(OF WHICH SATURATED FATS	3.4g
OF WHICH MONO-UNSATURATED FATS	3.7g
OF WHICH POLYUNSATURATED FATS)	0.6g
CARBOHYDRATE	12.5g
(OF WHICH ADDED SUGAR)	NIL
DIETARY FIBRE	0.4g
SALT	0.6g

Courgette Bake – Ready Meal (454g) Serves 2

Co-Op Stores

C.W.S. Technical Group, 28, Knowsley Street, Manchester, M8 8JU.

The Co-Op has recently launched a campaign called 'Eat Right Eat Well'. It includes a personalised meal plan, printed out by a computer to take away. There is also a help line for customers with nutritional queries (Tel 061-832-5464). A range of leaflets are available in Co-Op stores. No decisions have been made yet about the additive issue.

CONSUMER CARE

NUTRITION INFORMATION

100 GRAMS OF THIS DRAINED SWEETCORN TYPICALLY PROVIDES

2.4 grams of protein	LOW
20.0 grams of carbohydrate	HIGH
0.8 grams of fat	LOW
Energy value	390 kJ
(Calories)	(92 kcal)

PRODUCE OF USA

INGREDIENTS: Sweet Corn. Water. Sugar. Salt.

STORAGE INSTRUCTIONS

Although this product is packed in a lead free can with a protective white lining, the unused contents of all canned foods should be removed from the can, placed in an alternative covered container and kept in a refrigerator.

CO-OPERATIVE WHOLESALE SOCIETY LTD. NEW CENTURY HOUSE, MANCHESTER

DIRECTIONS

Serve cold corn straight from the can or empty contents into a saucepan and heat. Once hot, drain, add a pat of butter and serve.

GUARANTEE

Your Co-op will honour all your rights as a Consumer. If you have any reasonable complaint about this product, please return it to the Co-op where you bought it, and it will be replaced or your money refunded, as you wish.

People who care

Sweet Corn (326g)

Marks and Spencer
Michael House, Baker Street, London, W1A 1DN.
Marks and Spencer say their policy is to sell wholesome palatable food. They try to minimise the number of additives and avoid where possible preservatives, artificial colours and flavours. Additives needed for safety will continue to be used and so too will colours where necessary but colourings will increasingly be natural substances.

They have begun to use nutritional labelling on some products such as low-fat yoghurts. The Customer Services Department will issue lists of Marks and Spencer foods suitable for those on special diets and for anyone interested in avoiding specific additives. Write to them at the above address.

Presto
Argyll Stores, Argyll House, Millington Road, Hayes, Middlesex.
No official policy on additives yet. In-store leaflets available from the Customer Service desk in Presto Food Markets, including diet plan leaflets for healthy eating.

Waitrose
John Lewis Partnership, 4, Old Cavendish Street, London, W1A 1EX.
Waitrose have been reviewing their own brand products over the past two years. They aim to remove any additive that is not strictly necessary and use natural alternatives where possible. They have removed Brown FK from kippers and replaced it with 'Annatto', a natural substance. Their smoked cod and haddock, which used to contain Tartrazine and/or Sunset Yellow, now has 'Crocin', another natural substance. Already there are over 300 products free from artificial additives.

Lists of Waitrose own label products that are suitable for people on special diets, e.g. lactose-free, suitable for vegans, and free from artificial colours, preservatives and flavourings, etc. are available.

Asda Stores
Asda House, Britania Road, Morley, Leeds, LS27 0BT.
Asda say they have a policy of ensuring that the use of additives is kept to a minimum, whilst bearing in mind that there are certain features of additives which are beneficial to the shopping public in terms of shelf life etc. Limited own brand range.

Chapter Two
The Law and Labels

The Label
The labels on food packaging do give a fair amount of information about the food contained within it. All foods have to be marked with the name of the food, a list of ingredients, some indication of how long it may be stored ('Sell by' or 'Best before' dates) and how to store if special conditions are needed (e.g. 'store in a cool place' or 'store in a dry place'). In addition it should state the name and address of the manufacturers, seller or packer, and the place of origin (under certain circumstances). Instructions for use are also included for some foods.

Name of the Food
This sounds simple enough but given the flights of fancy that some advertisers come up with when involved in naming a new product, the law wisely insists that the name used for the food should indicate the true nature of the food, and enable it to be distinguished from products with which it could be confused. Trade marks, brand names and fancy names can be used in addition to the descriptive name. So a packet labelled 'Dreamy Fluff' would also have to give an indication of what it was, such as 'Fruit Dessert'.

Names should not be misleading. However very small changes in the name can make a great deal of difference. 'Apricot' or 'Apricot flavour*ed*' written on the packet or container shows that the flavour comes mainly from real apricots. 'Apricot flavour' however means just that it

WINDMILL BAKERY
800g Family Wholemeal Bread

INGREDIENTS

Wholemeal Flour, Water, Yeast, Salt, Hydrogenated Vegetable Oil, Emulsifier E472(e) (improves the eating qualities of the bread), Dried Glucose Syrup, Preservative E280, E281 (inhibits mould during warm weather), Flour Improver (E300).

THIS LOAF CONTAINS NO ANIMAL FATS

For 'sell-by' date see bag closure or label.
Best eaten within two days of 'sell-by' date, except in particularly warm weather, when we recommend that you eat it within one day. Always store in a cool place.

NUTRITIONAL INFORMATION		
Typical Analysis	Per 100g	Per daily serving of 4 slices (130g)
Energy	910kJ (215kcal)	1180kJ (280kcal)
Protein	9.5g	12.4g
Carbohydrate	41.2g	53.6g
Fat	2.5g	3.3g
Dietary Fibre	8.6g	11.2g
Salt	1.3g	1.7g

Family Wholemeal Loaf (British Bakeries Ltd.)

---

tastes of apricots but the flavour need not come from apricots. Check carefully when buying to see the exact description given. Also look at the picture on the packet or container. If the food in question has a picture of the fresh fruit on the label it is more likely to contain the fruit. The law says the label must not mislead the customer and can only show the fruit on the label if it contains that fruit.

Meat that has previously been frozen but is sold thawed *should* have a label saying 'Previously frozen – do not refreeze'.

The total weight or quantity of a product should be shown too, or the total number in the case of some breakfast cereals, eggs and other foods sold by number.

The other important aspect is an indication of what form the food is in. With so many packets on the supermarket shelves an indication of whether the food is powdered, dried, freeze-dried, frozen, concentrated or smoked is essential.

A description of how to use the food is sometimes needed along with instructions for use where necessary. The packet must also list the extra ingredients needed apart from water. So a cake mix must say 'Add butter and one egg' or words to that effect. But a dried soup does not need to say 'Add water' on the label.

Ingredients

The list of ingredients on many packages makes interesting reading. While at first glance they may look like a chemist's shopping list, they do give a great deal of information if you know how to interpret them.

First and foremost the ingredients have to be listed in descending order of weight. So you can generally tell how much sugar there is in your packet of biscuits or can of baked beans by seeing how high up the label it is. Sugar is sucrose, but other sugars include glucose, fructose and glucose syrup. Make a mental note where sugars are listed as several different types of sugar. Almost all sugars are detrimental to teeth and the total of the different types of sugars may be much greater than it seems from the label.

When food is sold in a dried or concentrated form the ingredients may be listed in the ratios they will be once it is reconstituted, as long as there is an indication to this effect. Usually the words 'Ingredients of the reconstituted product' or 'Ingredients of the ready-to-use product' are shown.

Some foods which consist of a variable mixture of different foods can be labelled as such. So a label may say 'Fruit and nuts in variable proportions'. Sometimes part of an ingredients list will have this heading. This shows that some components of a food consist of a variable mixture of certain foods. Foods likely to be labelled like this include mixed fruits, nuts, vegetables, spices and herbs.

The names of the ingredients, generally speaking, have to be the names which the food could be sold under separately, e.g. flour, sugar, carrots, vinegar, etc. Some names or descriptions need more specific legislation. For instance, the names of fish. Common names such as

Huss, Eel or John Dory refers to the specific species of fish — all of which are listed in the regulations covering food labelling. Melons and potatoes must be labelled with the species, e.g. King Edward potatoes or Honeydew melons.

Sometimes foods may be described by a general name such as flour or nuts. Where these are used there are sometimes conditions on their use. So, for instance, a list of ingredients may include the word 'fat' but should also include the description 'animal' or 'vegetable' plus an indication of the specific animal or vegetable origin. The terms 'nuts' can be used to mean any type of nut as long as the proportion does not exceed 1%, or if they are being sold prepacked with muscatels, raisins, sultanas and currants in a pack weighing not more than 50g (1 oz). 'Vine fruits' can refer to any muscatels, raisins, sultanas or currants. 'Starch' means any starch other than the one that has been chemically treated to modify it when the term 'modified starch' must be used.

If flour is shown on a list of ingredients it must be followed by a list of the cereals from which the flour is derived in descending order of weight. Where the term 'Other meat' or 'Other fish, is used it can be any other meat/fish apart from that listed on the label. So you could buy a product that lists 'Pork and other meat' without any clear indication of which other meat is included in the food.

'Sell By' and 'Best Before' dates

'Sell by' dates were introduced some 10 years ago amid much discussion and reluctance on the part of some supermarket chains and manufacturers. They have worked very well, with the consumers being able to check how fresh food is and buy accordingly. They also help guide the consumer when storing food at home. Perishable foods intended to be used within six weeks have to have a 'Sell by' date marked on them followed by the day and month they should be sold in. In addition they should indicate how long after purchase they can be kept and indicate any storage conditions needed. So a yoghurt or cream label needs to contain: 'Sell by 6th Jan; *and* 'Use within' or 'Best within 2 days of purchase' *and* 'Store in a cool place'.

Foods designed to keep longer than six weeks, i.e. most dry foods and canned foods as well as biscuits etc. should have the words **'Best before'** followed by the date and give any storage conditions needed to keep the food, e.g. 'Best before 6th Jan 1987'. If the food is best stored for less than three months the date may be expressed as day and month. Where the food is designed to store for more than three months the words 'best before the end of' followed by a month and year can be used instead.

Some foods need not be marked with 'Sell by' or 'Best before' dates. These include the following: fresh fruits and vegetables, wines and most alcoholic drinks, beer designed for resale, flour products and bread normally consumed within 24 hours of baking, vinegar, cooking salt, solid sugar or sugar products, frozen foods, edible ices, chewing gum and similar products, and cheese which is meant to ripen in the packaging. In addition to these any food that has a minimum life of 18 months does not need to be labelled. Its not against the law to sell foods after the datemark, as long as the customer realises the date mark has been passed and the food is still fit for human consumption.

Additives

Recent changes in the labelling of foods means that all additives have to be listed on the label giving either their name or the number and the purpose for which the additive may be used, e.g. **'E320 anti-oxidant'** or **'E102 colour'**.

Certain additives do not have to be listed, such as ingredients which have been temporarily separated during the manufacturing process and are later reintroduced in their original proportions. Any additive which is used solely as a processing aid can be omitted as can, any additive that is used solely as a solvent or carrier for an additive. Water usually has to be listed on the label of ingredients unless it is less then 5% of the finished product. Water used during the preparation of the food solely to reconstitute an ingredient used in a concentrated or dehydrated form does not have to be listed. So 'long life' apple juice in a carton may well have been concentrated for easy transportation and then reconstituted before being packed into cartons. Some foods need not have a list of ingredients on them: fresh fruit and vegetables (as long as they have not been cut or peeled), vinegar (as long as no other ingredient has been added), cheese, butter milk and cream, flavourings, foods consisting of a single ingredient, e.g. flour or sugar and most alcoholic drinks.

Similarily additives are not allowed in some foods. Fresh fruits and vegetables may not be coloured for instance and neither can milk, tea, coffee, fresh meat, poultry or fish.

Sometimes special ingredients are used in foods and promoted as making them extra special, e.g. real butter in biscuits, or with fresh cream in ice cream. Here the law lays down that a minimum percentage of that ingredient is prominently displayed next to the name of the food or in the list of ingredients near the special ingredient. Similarly if special emphasis is given to a low content of any ingredient, e.g. 'low fat', the actual amount in percentage terms must be included in the label. Of course, that sort of information is only really useful if you know roughly how much other versions of the same food have in them. If there is no indication on other food packs compare the ingredients list and see how high up the list that special ingredient comes. Remember, the higher up the list the greater the proportion.

Additive Categories

Some additives must be listed by their category name (which defines what they do in the food) followed by their specific name or number or both. Where there is no specific category it has to be identified by it's name. The categories are as follows:

Acids
Acidity Regulators
Anti-caking Agents
Anti-foaming Agents
Anti-oxidants
Artificial Sweeteners
Colours
Emulsifiers
Emulsifying Salts

Flavour Enhancers
Flavourings
Flour Improvers
Gelling Agents
Preservatives
Raising Agents
Stabilisers
Thickeners

The Exceptions
There are, as ever, exceptions to the rules: food which is not prepacked, and flour confectionery which is packed in transparent packaging and marked with just the price, flour confectionery, bread and ice creams are all excluded if they are made on the premises owned by that retailer. So bread baked in the supermarket's own bakery, for example, does not need the detailed labelling that most other foods have. They do, however, have to display a notice nearby to indicate that the items sold may contain additives. Very small packages are exempted too; any packet whose area is less than 10 square centimetres needs no list of ingredients. Food sold to be eaten immediately is generally exempt including hot meals, sandwiches and food sold in vending machines.

Another area where additives can go unlisted quite legally is that of the primary and secondary ingredients. If a manufacturer sells bread, he must list any additives in the flour. If, however, a manufacturer buys in flour and then makes a frozen pizza from it he can put flour on the list of ingredients without including any additives in it, because he did not add them. So occasionally the comment 'No added colouring or preservative' may simply mean that the final manufacturers did not add any. Look instead for the comment that the product contains no artificial colour/preservatives, etc.

Who Should Avoid Additives?

1. Hyperactive Children
Hyperactive children are often restless, need little sleep and can be very disruptive. They are often very thirsty and may suffer from headaches, asthma and eczema too. The Hyperactive Children's Support Group (HACSG) recommend that a diet based on the Fiengold diet is followed. This involves avoiding foods which contain synthetic colours, flavours, anti-oxidants and flavour enhancers. At the beginning of the diet some fruits and vegetables (those which contain natural salicylates) should also be avoided and then gradually reintroduced.

Since many of the foods sold for children are highly

coloured this means careful scanning of the packets and jars. However some supermarkets such as Safeway and Waitrose are gradually removing all these contentious additives and some also issue lists of their own brand foods free from artificial additives. See Chapter 1 for more information.

Hyperactive children should avoid:

E102	Tartrazine	155	Brown HT
E104	Quinoline Yellow	E180	Pigment Rubine
107	Yellow 2G	E210	Benzoic Acid
E110	Sunset Yellow	E211	Sodium Benzoate
E120	Cochineal	E320	Butylated Hydroxyanisole (BHA)
E122	Carmoisine		
E123	Amaranth		
E124	Ponceau 4R	E321	Butylated Hydroxytoluene (BHT)
E127	Erythrosine BS		
128	Red 2G		
E131	Patent Blue V	621	Sodium Hydrogen L-Glutamate (monosodium glutamate)
E132	Indigo Carmine		
133	Brilliant Blue FCF		
E142	Green S		
E150	Caramel	622	Potassium Hydrogen L-Glutamate
E151	Black PN		
E153	Carbon Black	623	Calcium Glutamate
154	Brown FK		

Detailed lists of additives are given in Chapter 3. This symbol indicates that a particular additive concerns those with hyperactive children who should be aware of foods which contain it.

2. Aspirin-Sensitive

Some people who are sensitive to aspirin are also sensitive to azo dyes and certain other synthetic additives. Not all aspirin-sensitive people will react badly to these substances, but if allergic-type symptoms (breathlessness, rashes, streaming eyes, etc.) occur it may be worth considering avoiding these additives. Consult a doctor and check if there are more obvious causes as well.

E102	Tartrazine	133	Brilliant Blue FCF
E104	Quinoline Yellow	E142	Green S
107	Yellow 2G	E151	Black PN
E110	Sunset Yellow	154	Brown FK
E120	Cochineal	155	Brown HT
E122	Carmoisine	E180	Pigment Rubine
E123	Amaranth	E212–E219	Benzoates
E124	Ponceau 4R	E220–E227	Sulphites
128	Red 2G	E312	Dodecyl Gallate
E132	Indigo Carmine		

Detailed lists of additives are given in Chapter 3. This symbol indicates that a particular additive concerns those who are sensitive to aspirin, who should be aware of foods which contain it.

3. Asthmatics

Asthmatic people can be sensitive to many things both natural and artificial. It may be worth avoiding some additives in food to minimise the effects of asthma.

E102	Tartrazine	133	Brilliant Blue FCF
E104	Quinoline Yellow	E142	Green S
107	Yellow 2G	E151	Black PN
E110	Sunset Yellow	154	Brown FK
E120	Cochineal	155	Brown HT
E122	Carmoisine	E180	Pigment Rubine
E123	Amaranth	E212–E219	Benzoates
E124	Ponceau 4R	E220–E227	Sulphites
128	Red 2G	E312	Dodecyl Gallate
E132	Indigo Carmine		

Detailed lists of additives are given in Chapter 3. This symbol indicates that a particular additive concerns asthmatics, who should be aware of foods which contain it.

4. Light Sensitive

Certain individuals are sensitive to light and should avoid these additives which can cause problems.

E131	Patent Blue V	E239	Hexamine
E132	Indigo Carmine	E280–E283	Propionic Acid and Propionates
133	Brilliant Blue FCF		
E142	Green S	E420	Sorbitol
E151	Black PN	E421	Mannitol

Detailed lists of additives are given in Chapter 3. This symbol indicates that a particular additive concerns those who are sensitive to light, who should be aware of foods which contain it.

5. Doubt Expressed about Safety

Many additives have been implicated in health problems. Some cause allergies, rashes, streaming eyes and upset stomachs. Some are implicated in more major problems like kidney stones and ulcerative colitis. Some are also thought to contribute to cancer. These are well worth avoiding where possible.

E123	Amaranth	E407	Carrageenan
E127	Erythrosine BS	430	Polyoxyethylene compounds
128	Red 2G	435	
E132	Indigo Carmine	E466	Sodium Carboxymenthyl-cellulose
133	Brilliant Blue FCF		
E150	Caramel		
E151	Black PN	507–510	Chlorides
154	Brown FK	553	Silicates
E230	Biphenyl	621	Sodium Hydrogen L-Glutamate (monosodium glutamate)
E239	Hexamine		
E249–E252	Nitrates and Nitrites		
E320	Butylated Hydroxanisole (BHA)	631	Inosiate
		907	Refined Microcrystallin Wax
E321	Butylated Hydroxytolulene (BHT)	924	Potassium Bromate

Detailed lists of additives are given in Chapter 3. This symbol indicates that a particular additive has had doubts expressed about its safety.

6. Babies and Children
Some foods contain additives which can affect babies and young children. These should be avoided.

E249	Potassium Nitrite	E325	Sodium Lactate
E280–283	Propionic Acid and Propionates	E326	Potassium Lactate
		E327	Calcium Lactate
E310	Propyl Gallate	514	Sodium Sulphate
E311	Octyl Gallate	515	Potassium Sulphate
E312	Dodecyl Gallate	621	Sodium Hydrogen L Glutamate
E320	Butylated Hydroxyanisole (BHA)	622	Potassium Hydrogen L-Glutamate (Mono potassium glutamate)
E321	Butylated Hydroxytolulene (BHT)	623	Calcium Glutamate
		631	Inosiate

Detailed lists of additives are given in Chapter 3. This symbol indicates that a particular additive concerns those with babies and young children, who should be aware of foods which contain it.

7. Vegetarians
A few additives are derived from animals. Vegetarians might wish to avoid foods containing these.

E120	Cochineal	542	Edible Bone Phosphate

Detailed lists of additives are given in Chapter 3. This symbol indicates that a particular additive concerns vegetarians, who should be aware of foods which contain it.

8. Vegans
In addition to those additives mentioned in the section above, vegans may wish to avoid any additive which is derived from animal produce.

E270	Lactic Acid	E327	Calcium Lactate
E325	Sodium Lactate	901	Beeswax
E326	Potassium Lactate		

Detailed lists of additives are given in Chapter 3. This symbol indicates that a particular additive concerns vegans, who should be aware of foods which contain it.

Chapter Three
Additives

The number of additives in our food has undoubtedly grown rapidly. We are all eating more processed and packaged foods, and less fresh homemade dishes. So we are eating more of the additives that are put in foods to make them last longer on the supermarket shelves, and help them to look brighter and more colourful.

The safety of certain additives for certain people is in question. Where there is any doubt it seems wise to avoid the use of the additive where possible. Gradually some manufacturers and supermarket chains are becoming aware of the resistance to the foods with additives but meanwhile it is up to the customer to read the label and vote with the purse, by choosing products free from the contentious additives.

The E numbers mean that the substances have been approved under an EEC directive. However, some of these additives are banned in other countries while still being used here. Some do not have an E prefix. This means either that the substance is proposed but not yet accepted, or that it has been turned down for an E rating. In each case there is no way of knowing which applies. Gaps in the numbers show just how many substances that did have an E number have now been dropped.

Being on the approved list seems to confer a feeling that they must be safe. But the whole area of safety of additives is a controversial one. Opinions vary greatly as to which additives are dangerous and to whom. There is also a cocktail effect. One additive may be safe but how about when mixed with all the other ones the average person consumes in a day or week or year? Safety is also based on average daily doses, in average portions. No allowance is made for people who eat a great deal of any one particular food. Anyone with a child hooked on crisps or addicted to hamburgers will know how difficult it can be to control the diet of children. While eating in this manner is not good for anyone it does happen. Then the idea of average daily doses is of little or no use.

The other side of the coin is that some additives are needed for safety. The preservatives in particular are used to prevent bacteria growing and producing toxins which are poisonous. Remove the preservative and you risk making the food dangerous itself. Strict rotation of food stocks, limited shelf lives and careful temperature control are needed to handle foods with fewer preservatives. Twenty years ago when additives began to be used on a large scale many shops would not be able to supply those needs. Now, however, even local corner shops have chilled display cabinets, and most supermarkets can control rotation of stock and temperatures of storage units easily. Some even have computerised re-ordering which keeps a constant tally of what goods are sold each day in each supermarket and re-orders stock. The technology is available to control food supplies properly and so preservatives should be less necessary than they were a couple of decades ago.

Most shoppers soon become accustomed to what the E numbers mean and which they wish to avoid. Listed here are all the E numbers and their chemical names. These do not represent all additives in food; far from it; only about a tenth of the additives used are controlled by specific legislation. Most flavours and sweeteners are not covered yet they represent a large number of the additives used. Saccharin, an artificial sweetener, carries a health warning in America!

Note: In the listings that follow a system of symbols has been introduced to allow you to identify quickly and easily those additives which may have side-effects for certain individuals. Use these listings and the lists that follow to identify the additives that are of concern to you and in which foods they are most likely to be found.

H Avoid for hyperactive children

a Avoid for aspirin-sensitive individuals

A Avoid for asthmatics

L Avoid for light-sensitive individuals

? Doubts about safety

C Avoid for babies and children

V Avoid for vegetarians

v Avoid for vegans

Colours – the 100's

The colour of food matters a great deal. 'We eat with our eyes' as the saying goes and tend to expect certain foods to be certain colours. This is no problem with fresh raw foods, but processing, cooking and preserving foods affect their appearance. The manufacturers then put back the colours, using either natural or synthetic dyes.

Natural or nature identical colours offer few problems. The synthetic ones however are a major cause of worry about additives in foods. The azo dyes are the main culprit. Many people sensitive to asprin are also sensitive to the azo dyes and they affect asthmatics and some people with eczema too. The Hyperactive Children's Support Group include the azo dyes on their list of additives to be avoided, along with other synthetic dyes.

There is quite a wide choice of colours available and some supermarket chains and manufacturers are finding it possible to gradually drop the use of these dyes and replace them with the natural ones or in some cases leave the food uncoloured. The French seem to like their *petit pois* a dull murky green in the can and it certainly does not affect the flavour. Perhaps a combination of safer dyes and more realistic expectations of the colours of processed foods will lead to less use of colours.

The 100's – Colours

E100 Curcumin
Extracted from the spice turmeric. Natural orange colour. Used in sauces, soups and some pastry dishes.

E101 Riboflavin
(can also be called Lacto flavin)
B group vitamin, usually produced from yeast. Nature indentical form also produced. Gives a yellow colour to cheese.

E102 Tartrazine
Used very widely in many processed foods including soft drinks, biscuits, cakes, convenience foods, sweets, sauces such as salad cream, and smoked fish. Yellow colour. Avoidance recommended by HACSG. Avoid for asthmatics and aspirin-sensitive individuals. Banned in some countries.

E104 Quinoline Yellow
A synthetic coal tar dye. Used for its green/yellow colour. Avoidance recommended by HACSG. Also safety of these dyes is in question. Avoid for asthmatics and aspirin-sensitive individuals.

107 Yellow 2G
A synthetic coal tar dye. Avoidance recommended by HACSG. Avoid for asthmatics and aspirin-sensitive people.

E110 Sunset Yellow
(can also be called Orange Yellow)
A synthetic coal tar dye. Used in a wide range of processed and convenience foods. Avoidance recommended by HACSG. Gives yellow/orange colour. Often used with colours such as E102. Avoid for aspirin-sensitive and asthmatic people. Banned in some countries.

E120 Cochineal
(can also be called Carmine of Cochineal or Carminic acid)
Natural red colour. Also a nature identical form is produced. May cause hyperactivity in children therefore avoidance recommended by HACSG, and for asthmatics and aspirin-sensitive individuals. Banned in some countries.

E122 Carmoisine
(also labelled as Azorubine)
Synthetic coal tar dye. Used to give a red colour. Found in jams, sauces and some breadcrumbs on prepared foods. Avoidance recommended by HACSG, and for aspirin-sensitive and asthmatic people. Banned in some countries.

E123 Amaranth
Synthetic coal tar dye. Used to give a red colour. Avoidance recommended by HACSG. Found in some bottled drinks, beefburgers and sauce mixes. Avoid for aspirin-sensitive individuals and for asthmatics. Banned in some countries including USA. Possible carcinogen.

E124 Ponceau 4R
(also known as Cochineal Red A)
Banned in some countries including USA. Synthetic coal tar dye. Used in jams, jellies, soups, some meat products and some tinned fruit. Gives a red colour. Avoidance recommended by HACSG for asthmatics and for anyone sensitive to aspirin.

E127 Erythrosine BS
Synthetic coal tar dye, so avoidance recommended by HACSG. Often found in prepacked ham and pork, and in flavoured crisps as well as sweet products like flavoured yoghurts. Possible carcinogen.

128 Red 2G
Azo dye, synthetically made. Gives an orange colour. Avoidance recommended by HACSG for aspirin-sensitive people and for asthmatics. Possible carcinogen.

E131 Patent Blue V
Synthetic dye. Gives a blue colour. Avoidance recommended by HACSG and for anyone who suffers from allergies.

E132 Indigo Carmine
(also known as Indigotine)
Synthetically produced dye used to give a blue colour. Avoidance recommended for hyperactive children. Used in sauce mixes and in meat products. Anyone suffering from allergies should avoid this as should asthmatics, aspirin-sensitive individuals and people suffering from eczema.

133 Brilliant Blue FCF
Synthetically produced dye used to give a blue colour. Avoidance recommended for hyperactive children, asthmatics, aspirin-sensitive people and those suffering from eczema. Banned in some countries. Possible carcinogen.

E140 Chlorophyll
E141 Copper complexes of Chlorophyll
Natural colouring agent. Nature identical form also produced. Banned in USA.

E142 Green S
(also known as Acid Brilliant Green BS or Lissamine Green)
Synthetically produced dye. Used in some tinned vegetables, some soups and drinks as well as jellies. Avoid for hyperactive children, asthmatics and aspirin-sensitive people and anyone suffering from eczema.

E150 Caramel
Brown colour produced by burning sugar or treating it with chemicals. The most widely used colouring agent. Used in a wide variety of foods and drinks. Also in some bread, cakes, biscuits, etc. Natural origin but doubts about its use and effects on B Vitamins in food have been voiced. Avoid for hyperactive children. Possible carcinogen.

E151 Black PN
(also known as Brilliant Black BN)
Synthetically made. Avoid for asthmatics, hyperactive children and for anyone who is aspirin-sensitive. Best to avoid for anyone with eczema.

E153 Carbon Black
(also known as Vegetable Carbon)
Of natural origin, but avoid for hyperactive children. Used in jellies, preserves and some fruit juices.

154 Brown FK
(sometimes called Kipper Brown)
Possible carcinogen. Synthetic dye banned in some countries. Used to colour fish such as kippers and some other foods. Avoid for hyperactive children, asthmatics and aspirin-sensitive people.

155 Brown HT
(also known as Chocolate Brown HT)
Synthetic dye. Avoid for hyperactive children, aspirin-sensitive people and asthmatics.

E160 a to f and E161 a to g — Carotenoids
Naturally occurring colours extracted from plants such as carrots, tomatoes, beetroot, apricots, etc.

E162 Beetroot Red
Natural plant pigment.

E163 Anthocyanins
Natural plant pigment.

E170 Calcium Carbonate
Naturally occurring. Used as colour but also for other purposes, e.g. acidity regulator. Much used in bread.

E171 Titanium Dioxide
Safe.

E172 Iron Oxides, Iron Hydroxides
Safe.

E173 Aluminium
Safe amounts likely to be found in foods as it is only used for decoration and to coat some pills.

E174 Silver
Safe in the quantities found in food. Used only in confectionery and decoration.

E175 Gold
Safe in the quantities found in food. Used only in confectionery and decoration.

E180 Pigment Rubine
(also known as Lithol Rubine BK)
Synthetic dye. Avoid for hyperactive children, asthmatic and aspirin-sensitive people.

Preservatives – the 200's

Foods can be preserved by many means. Our ancestors smoked, pickled, dried and salted foods to keep them through the winter. We use the traditional methods and add to those canning, freezing, sterilising and the use of chemical preservatives. Often preservatives are used in addition to other methods to increase the time they can be stored by preventing the growth of fungi, mould and bacteria.

Preservatives have been widely used as they allow the foods to be stored longer. This makes it possible to hold stocks in the warehouses and shops and supermarkets, which minimises losses from foods going 'off' between production and final consumption. Public concern about the amount of additives in foods has certainly been taken to heart by some companies and supermarkets. New lines of food and convenience meals are being produced with shorter shelf lives and fewer preservatives. As one food technologist put it 'The secret is going to be in temperature control in the future'.

It is worth remembering though that while some preservatives may be unwelcome, so too is the deterioration of food and the growth of bacteria, mould and other organisms that may well cause food poisoning. In some cases too the preservative contributes to the very nature of the food; bacon and hams, for instance, which without the controversial nitrates and nitrates would simply not be bacon and ham. The solution here seems to be to eat these foods in moderation. Make them occasional additions to the diet rather than a regular feature of it.

The 200's – Preservatives

E200 Sorbic Acid
Salts of Sorbic Acid E201, E202, E203
Natural or nature identical. May be used in soft drinks, cakes and some frozen foods.

E210–E219 Benzoates and complex Benzoates

Do occur in nature but most synthesised. Used in soft drinks, salad dressings and fruit pies. E210 and E211 should be avoided by hyperactive children. E212 to E219 should be avoided by anyone sensitive to aspirin and asthmatics.

E220–E227 Sulphites, Sulphur Dioxide

Synthetically produced and used in soft drinks (E220—Sulphur Dioxide) and in meat products (E223 Sodium Metabisulphite). E222 (Sodium Bisulphite) is sometimes added to wine and beer. Avoid if asthmatic or aspirin-sensitive. Banned in Japan.

E230 Biphenyl

(also known as Diphenyl)
Possible carcinogen.

E231 2-Hydroxybiphenyl

(also known as Orthophenylphenol)

E232 Sodium Biphenyl-2-yl-oxide

(also known as Sodium Orthophenylphenate)
Synthetic preservatives used to prevent fungal growths on the skins of citrus fruits. May penetrate into the fruit in some instances.

E233 Thiabendazole

Used as a preservative on banana skins and on some citrus fruit skins.

234 Nisin

Used as a preservative in some canned foods.

E239 Hexamine

Synthetic fungicide. May cause stomach upsets.

E249–E252 Nitrates and Nitrites
Of natural origins. Some such as Potassium Nitrate (E252) have been used for centuries in the preservation of meats. There is controversy about the possible carcinogenic effects of nitrates in the body. E249 Potassium Nitrite should not be included in foods intended for babies under 6 months. Use restricted in many countries.

E260–E263 Acetates
Of natural origins. Used in beer and vinegar, sauces and creams, bread, some crisps and some convenience foods.

E270 Lactic Acid
Occurs naturally in sour milk. Used in many foods both as a preservative and to balance the acidity and flavour foods. Found in cakes, breads, biscuits and some salad dressings.

E280–E283 Propionic Acid and Propionates
Naturally occurring substances. Avoid for migraine sufferers. Anti-fungal and mould-inhibiting agents. Banned in West Germany. Not allowed in baby foods in Britain.

E290 Carbon Dioxide
Natural origin. Used for cooling and for packaging as well as preserving.

296 Malic Acid
Naturally occurring and produced synthetically. Used in drinks, soups and sauces.

297 Fumeric Acid
Naturally occurring substance. Used in baked goods as an anti-oxidant.

Anti-oxidants and Acid Regulators – the 300's

Fats and oils tend to go rancid, and develop 'off' flavours. Anti-oxidants help prevent this happening. Many occur in nature and several vitamins have anti-oxidant effects including vitamin C and vitamin E.

Some of the synthetically produced anti-oxidants have side-effects for many people. Those susceptible to asthma and aspirin-sensitive people should avoid them. So too should hyperactive children. The use of these synthetic anti-oxidants seems to be quite variable. Compare labels and you will soon notice that they are used on some brands and not others, even though the food in the packet is apparently the same. So shop selectively to avoid the synthetic anti-oxidants as much as possible.

Acid regulators are often used alongside anti-oxidants but they are not usually associated with health problems.

The 300's – Anti-oxidants and Acid Regulators

E300 L-Ascorbic Acid
Salts of Ascorbic Acid E301, E302, E304.
Vitamin C. Occurs naturally but also produced synthetically. Added to food for its nutritional uses, but also acts to preserve food, helps colours combine with the food and as an improver to flour.

E306–E309 Natural and Synthetic Tocopherols
Vitamin E. Used as an anti-oxidant as well as for its vitamin benefit in the diet.

E310 Propyl Gallate
E311 Octyl Gallate
E312 Dodecyl Gallate
All three are synthetic anti-oxidants. Avoid for asthmatics, aspirin-sensitive people and hyperactive children. Not allowed in foods for babies.

E320 Butylated Hydroxyanisole (BHA)
E321 Butylated Hydroxtolulene (BHT)
Both are synthetic anti-oxidants. Avoid for hyperactive children. Not permitted in foods for babies. Doubts about safety expressed; suggestions that they may deplete body stores of vitamin D, and possibly be carcinogenic. Compare brands. The use of these two anti-oxidants is variable, e.g. some supermarket crisps have none while some branded ones have both.

E322 Lecithins
Naturally occurring fat. Used in chocolate-based products, some margarines and confectionery.

E325 Sodium Lactate
E326 Pottassium Lactate
E327 Calcium Lactate
All salts of Lactic Acid. Used as acid regulators and humectants. Not suitable for foods for babies.

E330 Citric Acid
Salts of Citric Acid: E331, E332, E333
Naturally occurring, but prepared by fermentation for commercial use. Used as an anti-oxidant as well as a preservative.

E334 Tartaric Acid
Salts of Tartaric Acid: E335, E336, E337
Natural origin. Used in baking powder, confectionery, fizzy drinks and packet dessert mixes.

E338 Orthophosphoric Acid
(also known as Phosphoric Acid)

E339, E340, E341 (all phosphates)
Naturally occurring but mainly synthesised now. Used as acid regulators, as a flour improver and in conjunction with anti-oxidants. Used mostly in potato products,

baking powder, fizzy drinks and packet dessert mixes.

350–352 Malates
Used to maintain acidity levels and to assist raising agents.

353 Metataric Acid
Slows down the setting of dessert mixes and combines with traces of metals to prevent oxidation.

355 Adipic Acid
Raising agent.

363 Succin Acid

370 Heptonlactone

375 Nicotinic Acid (vitamin B)

380 and 381 Citrates
Used in some bread flours.

385 Calcium Disodium EDTA
Stabiliser.

Emulsifiers, Thickeners, Anti-Caking Agents – the 400's and 500's

Emulsifiers hold together two substances that would not normally mix well. When making mayonnaise at home you are making an emulsion. Commercial ice cream uses emulsifiers and so do many prepacked or powdered dessert mixes.

Often an additive is used for more than one purpose: it may emulsify, stabilise and thicken the mixture as well. Sometimes it is also needed to allow the mixture to be processed by the machinery and packaged easily. Additives in this category also include anti-caking agents which prevent powdered or dry goods forming lumps. Release agents help to prevent the food sticking to the packaging or the machinery during production.

The 400's/500's – Emulsifiers, Thickeners, Anti-Caking Agents, etc

E400 Alginic Acid
E401 Sodium Alginate
E402 Potassium Alginate
E403 Ammonium Alginate
E404 Calcium Alginate
The alginates are all naturally occurring substances found in seaweed. They are used as stabilisers, emulsifiers and gelling agents. Found in ice creams, puddings and desserts.

E406 Agar
Gum from seaweed. Used as a stabiliser and gelling agent. A vegetarian alternative to gelatine.

E407 Carrageenan
(also called Irish Moss)
Gum found in seaweed and used in ice cream and drinks. May cause ulcerative colitis. May be carcinogenic.

E410 Locust Bean Gum (Carob Gum)
Carob is an alternative to chocolate. The gum is used in drinks, confectionery and some other foods.

E412 Guar Gum
Natural product extracted from seaweed. Used in diabetic foods. Added to sauces, soups and ice creams.

E413 Tragaxanth Gum
Natural origin. Used as an emulsifier and thickener. Also used in icings.

E414 Gum Arabic (Acacia)
Natural gum used as an emulsifier, thickener and stabiliser.

E415 Xanthan Gum
Of natural origin. Used as a stabiliser, emulsifier and thickener.

E416 Karaya Gum
Natural gum but banned in most of EEC.

E420 Sorbitol, Sorbitol Syrup
Occurs naturally, but produced from glucose for commercial use. Used in diabetic confectionery and as a general sweetener. Has a laxative effect if eaten in large quantities.

E421 Mannitol
A sugar alcohol extracted from seaweed. Widely used in ice cream and desserts as an emulsifier, thickener and sweetener. Some individuals may be sensitive to it. Nausea and vomiting are possible effects.

E422 Glycerol
Used to prevent cakes and icing absorbing moisture.

E430–E431 Stearates
Emulsifiers. May produce skin allergies and may be a possible contributory factor in the development of kidney stones.

432–433, 435–436 Polyoxyethylenes
Emulsifiers and stabilisers produced from Sorbitol. Banned in rest of EEC. 435 may be carcinogenic.

E440 a and b Pectin and Aminated Pectin
Naturally occurring and chemically treated pectin, used in jams and jellies.

442 Ammonium Phosphatides
Stabiliser used in chocolate products.

E450 a to c Polyphosphates
Emulsifier used in some cheeses to retain moisture and so prevent them drying out. Also used in some cooked meats.

E460–E463, E465–E466 Celluloses
Bulking agents. Used in low-calorie foods and high-fibre breads. E466 suspected of being carcinogenic.

E470–476, 478 Fatty Acids
Used as emulsifiers, stabilisers and anti-caking agents.

E481 Sodium Stearoyl-2-lactylate
E482 Calcium Stearoyl-2-lactylate
E483 Stearyl Tratrate
Used as stabilisers and emulsifiers in products such as crisps and breads.

E491–495 Sorbitans
Emulsifiers and stabilisers. Used in foods such as cake mixes.

500–504 Carbonates
Used to regulate acidity. Found in custard powders, ice cream, etc.

507–510 Chlorides
Used as salt substitutes, and in some canned foods. Suggestions that some may affect the intestines. Anyone with liver or kidney complaints may be wise to avoid these.

513 Sulphuric Acid
Poisonous irritant. Used to prevent bacterial growth in wine.

514 Sodium Sulphate
Dilutant. Tends to raise sodium levels in the body. Avoid for children and babies.

515 Potassium Sulphate
Useful salt substitute. Avoid for children and babies.

516 Calcium Sulphate
Sequestrant. Firming agent.

518 Magnesium Sulphate
Better known as Epsom Salts!

524 Sodium Hydroxide
Firming agent, solvent for colouring. Toxic. Irritant.

526 Calcium Hydroxide
Neutralizing agent.

527 Ammonium Hydroxide
Solvent for colourings.

528 Magnesium Hydroxide
Safer substitute for 524, 525 and 527 as an alkali.

529 Calcium Oxide
Used as an anti-caking agent in many cocoa products.

530 Magnesium Oxide
Used as an anti-caking agent.

535 Sodium Ferrocyanide

536 Potassium Ferrocyanide

540 Di-Calcium Polyphosphate
Added to cereals. Raising agent.

541 Sodium Aluminium Phosphate
Raising Agent.

542 Edible Bone Phosphate
Vegetarians may wish to avoid this.

544 Calcium Phosphates
Used in processed cheeses.

545 Ammonium Polyphosphates
Used in frozen poultry. Retains water.

551 Silicon Dioxide (Silica)
Thickener and stabiliser. Found in some wines.

552 Calcium Silicate
553a Magnesium Silicate
553b Talc
554 Aluminium Sodium Silicate
556 Aluminium Calcium Silicate
558 Bentonite
559 Kaolin
570 Stearic Acid
572 Magnesium Stearate
Anti-caking and release agents. 553 suspected as being carcinogenic.

575 D-Glucono-i-5-Lactone (Glucono delta lactone)
576 Sodium Gluconate
577 Potassium Gluconate
Calcium Gluconate
Acids and sequestrants.

The 600's + – Flavour Enhancers

Flavour Enhancers: 620–635

620 L-Glutamic Acid
Naturally occurring. Used as a salt substitute and flavour enhancer.

621 Sodium Hydrogen L-Glutamate
Known as Monosodium Glutamate. Flavour enhancer, used widely in meat products and convenience meals. Not suitable for babies and should be avoided by hyperactive children. Excessive amounts can cause migraine, nausea and palpitations in some people.

622 Potassium Hydrogen L-Glutamate (Mono potassium glutamate)
Flavour enhancer and salt substitute. Not suitable for babies. Should be avoided by hyperactive children.

623 Calcium Glutamate
Flavour enhancer and salt substitute. Not suitable for food intended for babies. To be avoided by hyperactive children.

627 Sodium Guanylate
Flavour enhancer. Used in snack foods such as crisps. People suffering from gout and those on a low purine diet should avoid this.

631 Inosiate
Flavour enhancer. Avoid for babies, children and those people with gout.

635 Sodium 5-Ribonucleotide
Mixture of 627 and 631.

636 Matol
637 Ethyl Maltol
Flavourings to enhance sweet or fresh baked smell in food.

The 900's – miscellaneous

900 Dimethypolysiloxane
Anti-foaming agents found in jams and some juices.

901 Beeswax
Glazing and polishing agent.

903 Carnauba Wax
Glazing and polishing agent used on chocolates and sweets.

904 Shellac
Glazing agent used in cakes, sweets and fizzy drinks.

905 Mineral Hydrocarbons
Glazing agent and sealing agent. Used on dried fruits, confectionery and cheese rind.

907 Refined Microcrystalline Wax
Polishing and stiffening agent used in some tablets and on chewing gum. Suspected as being carcinogenic.

920 L-Cysteine Hydrochloride
Improving agent for flour.

924 Potassium Bromate
Used to bleach flour, which can cause the vitamin E content to be reduced. It can also cause nausea and diarrhoea in sensitive individuals.

925 Chlorine
Used to bleach flour, which can cause the loss of vitamin E.

926 Chlorine Dioxide
Bleaching agent and oxiding agent.

927 Azodicarbonamide
(also called Azoformamide)
Flour improver.

Where the additives are used

These lists are by no means definitive. They are given as a guide to the sorts of additives found in different types of foods, and based on a cross section of the foods found in supermarkets. Compare brands and types as they often vary considerably.

Note: In the lists that follow some additives are shown in bold type. This means that these can affect certain people – those sensitive to aspirin, hyperactive children, etc. For full details of any side-effects see the lists on page 50–58.

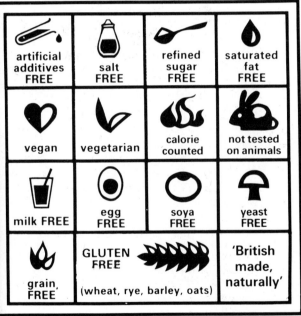

© *Rita Greer*

These symbols give a good deal of information about the additives in our food (see page 74).

Canned Foods

These include soups, vegetables, fruits and pie fillings, spaghetti, baked beans and meats. Most canned fruits are free from colours, with some exceptions, like cherries. Many fruits are now available canned in natural juices too. Fruit fillings intended to be used for pies and flans have many more additives, with colours and thickeners used quite widely. Canned meats in sauces often have thickeners added to the liquid, too.

- **E102 Tartrazine**
- **E124 Ponceau 4R**
- **E127 Erythrosine BS**
- **133 Brilliant Blue FCF**
- **E142 Green S**
- E160 Alpha, beta and gamma Carotene
- **E250 Sodium Nitrite**
- E301 Sodium L Ascorbate
- E330 Citric Acid
- E410 Carob Gum
- E412 Guar Gum
- E450 Sodium Polyphosphates
- **621 Sodium Hydrogen L-Glutamate (monosodium glutamate)**

Dried Foods

Dried foods often have emulsifiers and stabilisers in them and also ingredients to prevent them clogging together in the packet. Since many also have colour and flavour enhancers as well the list of additives can be quite long.

- E101 Riboflavin
- **E102 Tartrazine**
- **E110 Sunset Yellow**
- **E122 Carmoisine**
- **E123 Amaranth**
- **E124 Ponceau 4R**
- **E127 Erythrosine BS**
- **E150 Caramel**
- **154 Brown FK**
- E160 Alpha, beta and gamma Carotene
- E171 Titanium Dioxide
- **E180 Pigment Rubine**
- **E220 Sulphur Dioxide**

E222	**Sodium Hydrogen Sulphite**
E223	**Sodium Metabisulphite**
E262	Sodium Hydrogen Diacetate
E320	**Butylated Hydroxyanisole (BHA)**
E321	**Butylated Hydroxytoluene (BHT)**
E330	Citric Acid
E412	Guar Gum
E450	Trisodium Diphosphate
E466	**Carboxymethylcellulose**
E471	Mono and di-glycerides of fatty acids
E477	Propane-1, 2-diol esters of fatty acids
621	**Sodium Hydrogen L-Glutamate (monosodium glutamate)**
635	**Sodium 5-Ribonucleotide**

Cakes and Biscuits

Contents of these foods vary considerably. Many cakes are coloured and often creamy fillings for cakes and biscuits have a wide range of colours and flavours added to them. Some Swiss rolls and iced cakes have quite a collection of additives in them. We found one Swiss roll with four different colours (all of the azo dye group), one preservative and three different emulsifiers/stabilisers. In contrast some of the simpler biscuits have very few additives.

E102	**Tartrazine**
E110	**Sunset Yellow**
E122	**Carmoisine**
E123	**Amaranth**
E132	**Indigo Carmine**
E142	**Green S**
155	**Brown HT**
E160	Alpha, beta and gamma Carotene
E202	Potassium Sorbate
E270	Lactic Acid
E320	**Butylated Hydroxyanisole (BHA)**
E322	Lecithins
E330	Citric Acid
E331	Sodium Citrates
E440	Pectin
E461	Methylcellulose
E465	Ethymethylcellulose
E471	Mono and di-glycerides of fatty acids

E475 Polyglycerol esters of fatty acids
E477 Propane-l, 2-diol esters of fatty acids
E481 Sodium Stearoyl-2-lactylate
E500 Sodium Carbonate
504 Magnesium Carbonate

Drinks

Squashes, fizzy drinks and low-calorie drinks all tend to be high in colours and sweeteners. Many also have preservatives too. Alternative squashes are now available with fewer or in some cases no colourings. Fizzy mineral waters mixed with fruit juices make a good alternative fizzy drink for children.

E102 **Tartrazine**
E104 **Quinoline Yellow**
E110 **Sunset Yellow**
E123 **Armaranth**
E124 **Ponceau 4R**
133 **Brilliant Blue FCF**
E142 **Green S**
E211 **Sodium Benzoate**
E223 **Sodium Metabisulphite**
E304 Ascorbic Acid
E307 Synthetic Alpha-tocopherol
E320 **Butylated Hydroxyanisole (BHA)**
E331 Sodium Citrates
E414 Gum Arabic
E466 **Carboxymethylcellulose, sodium salt**

Snack Foods

Crisps and similar savoury snack foods are a useful alternative to sweets and chocolates for children. Some have synthetic flavours and colouring agents in them. Many are cooked in oils that contain anti-oxidants. However, there are several brands available which do not have anti-oxidants so shop around and compare labels.

E102 **Tartrazine**
E110 **Sunset Yellow**
E122 **Carmoisine**
E127 **Erythrosine BS**
E132 **Indigo Carmine**
E150 **Caramel**

E262	Sodium Hydrogen Diacetate
E320	**Butylated Hydroxyanisole (BHA)**
E321	**Butylated Hydroxytoluene (BHT)**
E322	Lecithins
E412	Guar Gum
E471	Mono and di-glycerides of fatty acids
621	**Sodium Hydrogen L-Glutamate (monosodium glutamate)**

Meat Products

Many meat products like pork pies, pasties, beefburgers, etc. contain preservatives. Some additives are used to improve the texture of processed meats too. Colouring agents like Red 2G are sometimes used as well to give a suitably pink colour.

E102	**Tartrazine**
128	**Red 2G**
E150	**Caramel**
E160b	Annatto
E223	**Sodium Metabisulphite**
E250	**Sodium Nitrite**
E321	**Butylated Hydroxytoluene (BHT)**
E332	Monopotassium Citrate
E400	Alginic Acid
E407	**Carrageenan**
E440	Pectin
E450a	Tetrasodium Diphosphate
E450c	Sodium Polyphosphates
E461	Methylcellulose
E471	Mono and di-glycerides of fatty acids
621	**Sodium Hydrogen L-Glutamate (monosodium glutamate)**
635	**Sodium 5-Ribonucleotide**

Ice Creams and Frozen Desserts

Many of these foods have high quantities of fat in them, and are whipped to aerate them. Because of this they need emulsifiers and stabilisers to maintain their texture. Many also have colouring agents added as well.

E100	Curcumin
E102	**Tartrazine**
E104	**Quinoline Yellow**

E110	**Sunset Yellow**
E122	**Carmoisine**
E123	**Armaranth**
E124	**Ponceau 4R**
E142	**Green S**
155	Brown HT
E211	**Sodium Benzoate**
E322	Lecithins
E331	Monosodium Citrate
E401	Sodium Alginate
E407	**Carrageenan**
E410	Carob Gum
E412	Guar Gum
E421	**Mannitol**
E465	Ethylmethylcellulose
E466	**Carboxymethylcellulose, sodium salt**
E471	Mono and di-glycerides of fatty acids
E475	Polyglycerol esters of fatty acids
E476	Polyglycerol Polyricinoleate

Preserves and Sauces

Many jams and marmalades have colours added to them and so do mustards and sauces. Peanut butter tends to need an emulsifier to maintain the texture and prevent the oil separating out. One brand now uses sea salt as a natural emulsifying agent in peanut butter, and sugar-free versions are also available.

E102	**Tartrazine**
E122	**Carmoisine**
E142	**Green S**
E150	**Caramel**
E151	**Black PN**
E171	Titanium Dioxide
E202	Potassium Sorbate
E330	Citric Acid
E331	Monosodium Citrate
E410	Carob Gum
E412	Guar Gum
E415	Xanthan Gum
E440a	Pectin
E471	Mono and di-glycerides of fatty acids

Breaded Products
Fish fingers led the revolution for quick cooking breaded foods and now a wide variety is available including chicken, turkey, ham and all sorts of fish in breadcrumbs. Colourings are often used in the breadcrumbs themselves and chemical dustings used under the crumbs. Preservatives are also often used on these products.

E102	**Tartrazine**
E104	**Quinoline Yellow**
E124	**Ponceau 4R**
E250	**Sodium Nitrite**
E251	**Sodium Nitrate**
621	Sodium Hydrogen L-glutamate (monosodium glutamate)

Dairy Products
Many yoghurts for children are highly coloured, but also there are now many brands without additives as well. Most margarines contain stabilisers and some have colours in. Many use natural colours rather than synthetic ones. Cottage cheese and cheese spreads tend to use stabilisers and emulsifiers.

E123	**Amaranth**
E124	**Ponceau 4R**
E160a	Alpha, beta, gamma carotene
E160b	Annatto
E202	Potassium Sorbate
E211	**Sodium Benzoate**
E270	Lactic Acid
E322	Lecithins
E334	Tartaric Acid
E410	Carob Gum
E412	Guar Gum
E415	Xanthan Gum
E450	Tetrasodium Diphosphate
E471	Mono and di-glycerides of fatty acids

Ready Meals
Dried meals may well contain preservatives, occasionally colours and quite often stabilisers and anti-oxidants. Frozen meals may have some colours, occasionally thickeners and flavour enhancers. Many additive-free versions are now available as well.

E102	**Tartrazine**
E110	**Sunset Yellow**
E124	**Ponceau 4R**
E127	**Erythrosine BS**
E150	**Caramel**
E160a	Alpha, beta and gamma carotene
E160b	Annatto
E171	Titanium Dioxide
E220	**Sulphur Dioxide**
E223	**Sodium Metabisulphite**
E301	Sodium L-Ascorbate
E320	**Butylated Hydroxyanisole (BHA)**
E321	**Butylated Hydroxytoluene (BHT)**
E322	Lecithins
E330	Citric Acid
E401	Sodium Alginate
E471	Mono and di-glycerides of fatty acids
E472b	Lactic Acid, esters of mono and di-glycerides of fatty acids
621	**Sodium Hydrogen L-Glutamate (monosodium glutamate)**

Sweets and Confectionery
Colours are considered important in foods aimed mainly at young children. Unfortunately many of the colours are synthetic ones. Sweeteners are also used, plus occasionally additives, to control the thickness of foods such as chocolate and other ingredients to prevent some sweets drying out.

E102	**Tartrazine**
E110	**Sunset Yellow**
E123	**Amaranth**
E127	**Erythrosine BS**
E132	**Indigo Carmine**
E150	**Caramel**
E153	**Carbon Black**

E160a	Alpha, beta and gamma carotene
E171	Titanium Dioxide
E222	**Sodium Bisulphite**
E330	Citric Acid
E400	Alginic Acid
E422	Glycerol
E440a	Pectin

Bread

Flours can be treated with substances to bleach it and 'improve' it. Preservatives such as anti-fungal agents are also used to prolong the shelf life of bread. Loaves without additives are available but they are best eaten within 24 hours, rather than the 48 hours allowed for bread made with additives.

E280	**Propionic Acid**
E281	**Sodium Propionic**
E300	L-Ascorbic Acid
E471	Mono and di-glycerides of fatty acids
E472e	Acetic Acid esters of mono and di-glycerides of fatty acids
E481	Sodium Stearoyl-2-lactylate
920	L – Cysteine Hydrochloride
924	**Potassium Bromate**
926	Chlorine Dioxide
927	Azoformamide

Chapter Four
Nutritional Labelling

Reports that tell us to change our eating habits are all well and good. However, they assume that we know exactly what goes into our food and which foods contain which nutrients. In practice it can be dificult to be sure. Butter and oils are obviously high in fats, but how much fat does pâté have in it or how much is there in a cream cracker? Even where it is obvious e.g. with a bottle of oil, it is almost impossible to tell if the oil is high in saturated fats without some indication on the label. Nutritional labelling is essential to go hand in hand with attempts to improve the nation's diet. There are currently proposals for limited nutritional labelling of food which should come into force in 1988. These refer mostly to fat labelling which would, under these proposals, be compulsory. All foods will have to be marked with the fat content and the proportion of saturated fats present specified. So labels will have to say 'Fat 20g, of which saturates 10g'. The customer will then know how much fat there is in the food and how much of it is saturated fat. There will be exceptions of course; any food or drink containing less than 0.5% fat, fresh fruits and vegetables, herbs, spices and seasonings, alcoholic drinks, bread flour and cereals, and foods sold for immediate consumption will not require fat labelling.

Many bodies have called for compulsory nutritional labelling. However, arguments abound over how much information will be understood, so for the moment at least, only the labelling of fat levels is to become compulsory.

Any other information given will be voluntary.

The Government has issued draft guidelines to ensure that nutritional information given voluntarily will conform to a standard format. This should avoid confusion and allow easy comparison of similar foods.

First on the list will be fat, followed by protein, carbohydrate, energy, and salt (listed as sodium). Carbohydrates can be broken down to show sugars and starch separately. Vitamins, minerals and dietary fibre can also be listed.

So a full label will given the following information:-

Sodium Fat Protein Carbohydrate Dietary Fibre	– all listed in grams – value shown where there is more than 1%
Energy	– listed in kilojoules(Kj) and kilocalories (Kcal)
Vitamins and Minerals	– amount shown only where one serving contains more than 17% of the recommended daily amount

Very small amounts can be listed as 'trace' but everything else has to be listed in terms of amounts per 100 grams or per 100 ml.

Listing the fat content of certain foods is not straightforward. Meats for instance, especially minced meats, will vary so in some cases the fat content will be listed in categories. Minced meat, for instance, is likely to be subdivided into categories, each with a maximum fat content such as 15%, 25%, and 40%. Cakes will be subdivided into groups such as 'sponge cakes', 'fruit cakes', etc. and the content for that group will be the figure given. The same system will apply to joints and cuts of meat. The actual breakdown of the groups has not been finalised yet, but even using this grouping system it should help the shopper. How often one buys minced meat only to find that large quantities of fat appear during the cooking!

Saturated and Unsaturated Fats

The recommendation is to cut down on fat consumption and to eat less saturated fats and more polyunsaturated fats.

But which are which? Generally saturated fats are hard at room temperature and mostly (but not exclusively) from animal sources. Unsaturated fats are generally found in vegetable oils and are usually liquid at room temperature.

Source High in Saturated Fats

Meat Products   Lard
Dairy Produce   Suet
Block Margarines   Cocoa Butter
Coconut Oil

Sources High in Unsaturated Fats

Sunflower Oil   Olive Oil
Corn Oil   Soya Bean Oil
Most Nuts and Nut Oils

The Coronary Prevention Group is currently carrying out scientific tests on a wide range of commonly used foods, and will be issuing recommendations for statements on high, medium and low levels of nutrients in food products. These could then be used by all food manufacturers and retail outlets. The guidelines, and indeed current practice, is for manufacturers and retailers to calculate the levels of nutrients in the food themselves, and this can lead to a confusing degree of variation.

These guidelines are still at the proposal stage, so changes may still occur to the details. However once they become widely adopted they should make shopping for a healthy diet very much easier. For some of those symbols currently in use see Chapter 5.

A standard format will allow easy comparison of foods. The potential for pictorial representation has been rather unexplored though. A recent competition run by the Coronary Prevention Group and the BBC2 *Food and Drink* TV Programme highlighted the effectiveness of good symbols to illustrate the nutritional content of food. Their aim was to see if it was possible to come up with simple, easily understood symbols for saturated fats, sugars, salt and fibre. These symbols would give simple 'at a glance' information to be used alongside the more detailed information charts, not instead of.

This is an important point, since simple symbols such as these cannot by their nature, carry the full information. They would supply an interim level of information to encourage shoppers to think about the nutritional content of the food and then, perhaps, read the detailed list.

The competition received hundreds of entries, from designers, students, school children and the general public. The striking point was that approximately half the entries used very similar symbols. This shows how acceptable symbols of this kind are likely to be.

Presented here are a selection of the prize winning symbols.
First Prize Winner – Senior Category: Leigh Goddard
Third Prize Winner – Senior Category: Geoff Anderson
Fifth Prize Winner – Senior Category: Marlow Foods
Specially Commended – Senior Category: Janet Hildred

First Prize – Junior Category: Rupert Smith
Second Prize – Junior Category: G. Alcroft
                            Paul Winters

The winning entries have been shown to the Ministry of Agriculture, Fisheries and Food, and interest has been expressed by several major food companies as well.

FIRST PRIZE – Senior Category
Leigh Goddard
Graphic Design Student
Doncaster Institute of Higher Education

THIRD PRIZE – Senior Category
Geoff Anderson
Graphic Designer
London

FIFTH PRIZE – Senior Category
Marlow Foods Ltd.
Bucks

SPECIALLY COMMENDED – Senior Category
Janet Hildred
York

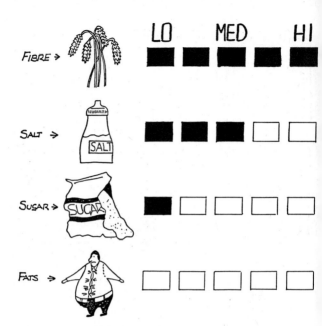

FIRST PRIZE – Junior Category
Rupert Smith
Leics.

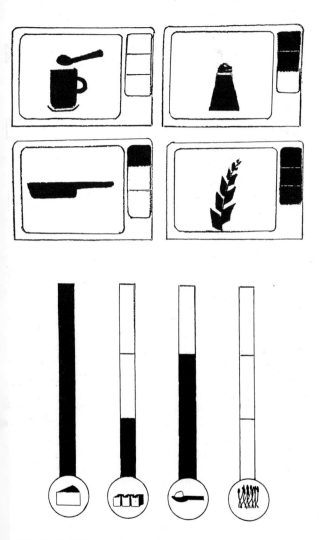

SECOND PRIZE – Junior Category
G. Alcroft
P. Winters

Chapter Five
Food Symbols

There is no standardization of symbols on food but they are gradually being used more widely. Presented here are those already in use that you are most likely to recognise:

Gluten-free symbol. Copyright by Coeliac Society. Widely used on baby food and on specialist foods available on prescription. Gluten is a protein found in wheat, rye, barley and, in some cases, oats. Gluten-free products use mainly maize starch for thickening.

Growing leaf symbol shows that food is suitable for vegetarians.

This symbol guarantees that the food on which it is displayed is organically grown. Awarded by the Soil Association which inspects and approves organically grown crops around the country.

Rita Greer Symbols

Designed by Rita Greer in 1977 and used on some diet foods and on vitamin tablets.

 Free from artificial additives.

 This symbol guarentees that the product is salt-free and will not add to your salt intake.

 Free from refined sugar.

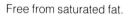 Free from saturated fat.

 This is the vegan symbol. Products displaying this are completely free from animal sources.

 The vegetarian symbol. Products bearing it are fully compatible with vegetarian ideals.

 The food displaying this symbol has been calorie counted.

 The milk-free symbol.

The egg-free symbol.

 The grain-free symbol.

 The soya-free symbol. More people are becoming sensitive to soya.

Not derived from yeast.

 Any product with this symbol on it indicates that no animal testing has been carried out by the manufacturers.

 This gluten-free symbol ensures the product is totally free from wheat, rye, barley and oats.

Chapter Six
Glossary of Food Terminology

Acid Regulators These maintain or change the acidity or alkalinity levels. Used for preservation. They help impart a tart flavour. Assist in the action of raising agents.

Anti-caking Agents These prevent powdered substances like salt, sugar and dried milk 'caking' or lumping together in the finished product.

Anti-oxidants Their main purpose is to stop the discolouration of fruit, vegetables and fruit juices. They also help prevent the rancidity of fats and so extend the shelf life of some foods. The main synthetic anti-oxidants are E310, E311, E312, E320 and E321. None of these should be used in foods for babies. E310, E311 and E312 should all be avoided by aspirin-sensitive people. E320 (BHA) and E321 (BHT) are both best avoided for hyperactive children. There are some doubts about the safety of these last two and they may affect vitamin levels in babies. The use of anti-oxidants is quite variable: some brands use them and some don't so scan the labels of crisps, margarine and dehydrated and convenience foods and choose those without anti-oxidants where possible.

Artificial Sweeteners These are sweetening agents and can be found in low-calorie products to aid slimming. Some such as Sorbitol have unwanted side-effects.

Aspartame The new artificial sweetener in products like *Candarel* and *Nutra Sweet*. It is from protein origin and has negligible calories. Used in confectionery and ice cream, low-calorie desserts and soft drinks.

Autolysed Yeast The yeast cells are ruptured to release their contents making the B Vitamins more readily available.

Buffers These maintain or change the acid or alkalinity levels. Used for preservation, they impart a tart flavour and assist in the action of raising agents.

Bleaching Agents These bleach or whiten flour and bread. Bleaching affects the vitamin content of flour. Some bleaching agents can irritate the digestive tract in susceptible individuals.

Bulking Agents These give extra bulk to food without adding extra calories. Used mainly in slimming and high-fibre products.

Caffeine A natural component of tea, coffee, cocoa and some drinks. Stimulant.

Calciferol/Cholecalciferol Vitamin D. Used to fortify margarine. It can be made naturally by the action of sunlight on the skin or extracted from yeast. Helps maintain healthy bones and teeth with the help of phophorous and calcium.

Colours Mainly used to restore food to its original colour after processing and to improve the appearance of bland or pale food. Many synthetic colours are worth avoiding for hyperactive children, asthmatics and aspirin-sensitive individuals. Often natural or nature identical colours are available as alternatives and more manufacturers are using these now. The synthetic colours are worth avoiding for everyone and these include E102 (Tartrazine), E104 (Quinoline Yellow), E109 (Yellow 2G), E110 (Sunset Yellow FCF), and E120, E122, E124, E127, 128, E131, 133, E142, E150, E151, E153, 154 (Brown Fk), 155 and E180.

Dilutents These dilute or dissolve other additives. Often used to help colours blend in and combine with the food.

Edible Starch Used to stabilise and modify the textures of sauces and chutneys. Usually from corn, potato and wheat. Beware if on a gluten-free diet.

Emulsifiers These are used to form an emulsion between two substances by suspending tiny particles of one liquid in the other. They also reduce the quantity of fat in foods and can be added to ice creams and cakes that normally require high levels of fat for production. Some emulsifiers are also used in baked products to make them last longer.

Extenders Used in meat products to make the meat go further. Often based on soya bean products which are a good source of protein.

Fibre This is a mixture of indigestible materials and is not absorbed into the body. Found in cereals, fruit and vegetables. Recent reports recommend that we increase our intake of fibre by 25%.

Firming Agents These help prevent foods from falling apart and softening when they are canned or frozen. Keeps fruit and vegetables firm and crisp.

Flavour Enhancers These restore flavours lost in processing. Some like 621 Monosodium Glutamate have side-effects (headaches and sickness) for many people. Widely used to heighten the taste of foods.

Flour Improvers These accelerate the bleaching process in white flour and help extend the elastic properties of dough in breadmaking.

Freezants They extract heat from food that is to be frozen.

Fructose Naturally occurs in some fruit and vegetables especially honey. It is the sweetest sugar known.

Glazing Agents Used to produce a sheen on the surface of the food and to form a protective coating.

Gelling Agents These give stability to food and improve texture.

Glucose Occurs naturally in fruit and plant juices and in the blood of living animals. Most carbohydrates are converted to glucose during digestion. It can be manufactured from starch by the action of acid on specific enzymes.

Glucose Syrup A mixture of maltose, glucose and other complex sugars. It is less sweet than pure glucose. Used in manufacture of confectionery and in some cakes and biscuits.

Humectants These retain moisture and stop the food drying out due to the evaporation of water.

Hydrogenated Vegetable Oil The oil is 'hardened' and is less likely to go bad when subjected to light, heat and traces of metals. Makes it more solid and more palatable.

Hydrolysed Vegetable Protein This can be chemically produced, or made by the enzyme breakdown of some raw protein materials. It is used to enhance the flavour of soups, gravy and meat products or to create a meat flavour.

Invert Sugar Made from glucose and fructose in equal amounts. Used as a sweetener and humectant. Also used to produce softer confectionery.

Iron This is needed to carry oxygen in the blood and in the formation of red blood cells. It is used to fortify breakfast cereals and added to white flour to compensate for the loss in processing.

Kilocalories This is a measurement of energy and is another name for calories.
1 Kilocalorie = 1 Calorie Usually abbreviated to Kcal.

Kilojoule This is the metric measurement of energy which must by law be shown on food labels with the kilocalories figure.
1 Kilocalorie = Approx. 4.2 Kilojoules
Usually abbreviated to Kj.

Malt Extract A medicinal body-building food made from malt.

Maltose Dextrin Carbohydrate, Used to change the texture, colour and flavour of the end product.

Modified Cornflour Almost pure starch with little nutritional value. Modified chemically to improve its ability to thicken and increase its shelf life.

Modified Starch Generally made from maize, potato and wheat. The starch is modified as the natural starch gels are unstable to heat and freeze. Added to sauces, desserts, meat, fish and dairy baked products. A general term for 18 substances not defined by E numbers. Plans were made to give them numbers (E1400–E1442) but were dropped and these numbers should not be used.

Monosodium Glutamate This is a flavour enhancer and works by stimulating the taste buds. It is used in foods containing protein. 621.

Nutrients These can either enrich foods by replacing vitamins and minerals lost during processing or they can be added to foods that may be lacking in essential nutrients needed in the diet, e.g. breakfast cereals, flour, etc.

Preservatives These extend the shelf life of products and may help to retain the natural colour or flavour of the food. Widely used though many are best avoided where possible.

Propellants Gases that are used to expel foods from aerosol containers.

Raising Agents These give a lighter texture and increase the volume of baked products.

Release Agents These stop foods sticking to surfaces during manufacture.

Polyunsaturated Fatty Acids Opt for polyunsaturated fats over saturated fats wherever possible. Choose vegetable oils such as sunflower and corn rather than blended oils for cooking.

Saturated Fatty Acids These are mainly from animal sources. An excess can lead to heart disease because a build up of fat in the arteries causes a strain on the heart. A 15% reduction in saturated fatty acids is recommended.

Sequestrants Slow down the deterioration of food and also slow down setting in dessert mixes.

Salt This gives flavour and acts as a preservative. Processed food is now the main contributor of salt to our diets. Look out for low-salt and salt-free foods. Many vegetables traditionally canned in brine are now available without salt.

Stabilisers Prevent emulsions separating out, thus improving the texture and consistency of the food.

Sucrose Occurs naturally in sugar cane and beet. Refined sugar is essentially pure sucrose. Recent reports suggest we should reduce sugar intake by 10%.

Thickeners These give stability to food and change the texture.

Vegetable Extract Usually a powder made from dried vegetable material. Added to improve the flavour, it is also a source of protein and minerals.

Yeast Extract This is a good source of some of the B vitamins. It may be useful to vegetarians as otherwise the sources are mainly animal foods.

Chapter Seven
Finding Out More

For anyone interested in the topic of what goes into our food and how healthy our national diet is, there are plenty of organisations around that offer more information and much of it is free. Some bodies will have a vested interest in the information they send out. Others are quite outspoken about certain topics, for instance, the dangers of additives.

Ministry of Agriculture, Fisheries and Food
Publications Unit, Lion House, Willowburn Trading Estate, Alnwick, Northumberland, NE66 2PF.
Leaflet available called *Look at the Label* which briefly gives information on the law regarding food labelling, and lists the E numbers and their chemical names, but does not give information about them.

The Vegetarian Society
Parkdale, Dunham Road, Altrincham, Cheshire.
This society has leaflets and books available on vegetarian cookery and nutrition. They also run cookery courses on healthy eating. They run a vegetarian centre and bookshop in London at:
53 Marloes Road, Kensington, London, W8 6LA.

Vegan Society Ltd.
33/35 George Street, Oxford, OX1 2AY.
Produce quarterly journals on vegan nutrition (i.e. a diet without any animal products including milk, eggs, cheeses, etc.) Also have book and leaflet list.

National Eczema Society
Tavistock House North, Tavistock Square, London, WC1H 95R.
Produce leaflets on diet and eczema including one called *Feeding Your Baby*. Local support groups.

London Food Commission
PO Box 291, London, N5 1DU.
A voluntary organisation involved in research, information, education and advice on food produce. Reports on many aspects of food and health, and also has a regular newspaper called *London Food News*. Send SAE for list of publications.

The Health Education Council
78 New Oxford Street, London, WC1A 1AH.
Range of leaflets available direct or from local sources such as libraries, doctors' surgeries and clinics.

Consumers Association
14 Buckingham Street, London, WC2N 6DS.
Lobbies on behalf of consumers of food issues. Also regularly covers food and diet in *Which* magazine.
Subscription department:
Consumers Association, PO Box 44, Hertford, SG14 1SH.

The Hyperactive Children's Support Group
c/o Sally Bunday, 59 Meadowside, Angmering, Sussex, BN16 4BW.
Advice and support for the parents of hyperactive children. Issue diet list and lists of food free from the additives they recommend hyperactive children to avoid. Organise local support groups. For information send SAE 9 × 4 inches.

The Soil Association
Walnut Tree Manor, Haughley, Stowmarket, Suffolk, IP14 3RS.
Produce useful leaflet called *Look Again At The Label*. Available free on receipt of SAE. Also produce book and leaflet list, and lists of farmers who grow crops organically.

Food and Drink Federation
6 Catherine Street, London, WC2B 5JJ.
Free, rather basic leaflet called *Food Additives*.

Coeliac Society of UK
PO Box 181, London, NW2 2QY.
Produce lists of gluten-free products available which is updated once a year. They have recipe books and leaflets on many aspects of coping with a gluten-free diet.

Action Against Allergy
43, The Downs, London, SW2 8HG.
Send a large SAE for publications list and information on additive-free foods.

The Coronary Prevention Group
60, Great Ormond Street, London, WC1N 3HR.
Tel: 01-833-3687
This is a broad based organisation that publishes leaflets and booklets on healthy eating in relation to heart disease. They will also help answer specific queries. Their leaflet *Healthy Eating and Your Heart* is available free. They also have a detailed publications list and a booklet called *Healthier Eating: A Good Foods Guide* which includes recipes.

The Asthma Society
300, Upper Street, London, N1 2TU.
Tel: 01-266-2260
Send a stamp for their free information pack which includes leaflets on asthma in the family, allergies and a publication list. Annual membership available (currently minimum of £2 a year) with 130 local branches that offer support and information.

Foresight – The Association for the Promotion of Pre-Conceptual Care
The Old Vicarage, Church Lane, Witley, Surrey, GU8 5PN.
Gives advice on nutrition and healthy eating prior to conception. Also aims to promote research aimed at identification and removal of potential health hazards to foetal development. Currently recommend a diet free from controversial additives where possible. Send an SAE for information to the above address.

Appendix

The tables in this section are designed as personal reference cards for each member of the family to fill in as regards those additives which may affect him or her. With this information available easily to hand the task of comparing your family's preferences with the food labels on the shelves is made much easier and you can be sure that the foodstuffs you choose are acceptable *before* you buy. Get into the habit of referring to your checklists and you will soon find you can remember which additives to avoid automatically. Also included is a space to note down any medication that is being taken as this can be important to bear in mind when considering which additives to avoid.

Additives Memo

Name:

Problem:

Additives/Foods to be avoided

Medication/Date taken

Additives Memo

Name:

Problem:

Additives/Foods to be avoided

Medication/Date taken

Additives Memo

Name:

Problem:

Additives/Foods to be avoided

Medication/Date taken

Additives Memo

Name:

Problem:

Additives/Foods to be avoided

Medication/Date taken

Additives Memo

Name:

Problem:

Additives/Foods to be avoided

Medication/Date taken

Additives Memo

Name:

Problem:

Additives/Foods to be avoided

Medication/Date taken

Notes